ROLLFORMING 101

How to choose the right technology

Dako KOLEV P.Eng.

Copyright 2017 – KOLEV Engineering Inc.

ISBN: 978-1-77277-133-6

IMPORTANT NOTICE

This book provides general guidelines in analyzing rollforming technology. In no case, however, are these guidelines intended to supersede any specific recommendations or instructions from the rollforming equipment manufacturer.

DISCLAIMER

The material in this book is intended for educational purposes and general information only. Any use of this material in relation to a specific application must be with the advice and professional expertise of an independent licensed professional engineer, specializing in rollforming equipment and tooling. Those making use of the material assume all risks and liability arising from such use.

PUBLISHED BY:
10-10-10 PUBLISHING
MARKHAM, ON
CANADA

This book is dedicated to those who are beginning their careers in the Roll Forming industry. It presents the basic Roll Forming principles in evaluating the roll forming process and developing awareness for potential mistakes in establishing a technology for specific roll formed products. This book will help you develop a better understanding of Rollforming Line Configuration, rolls tooling design, and trouble shooting.

This book is also dedicated to my girls: my wife Milena, and my daughters, Katerina and Veronica, who are always striving for nothing less than the best, and have always inspired me and supported my passion in designing and building machinery.

Table of Contents

Acknowledgments

Rollforming is a never-ending learning field with endless challenges. Working with people who have great technical philosophy, expertise and vision, and who are persistent in exploring the unknown, is a blessing I have had in my career.

First, I thank George Halmos, the author of *Roll Forming Handbook*, whose passion for roll forming guided me in my first steps in this field. Without reservations, he shared his deep and thorough knowledge with me, and coached me on how to avoid taking the wrong turns when attempting to solve specific roll forming challenges. He also helped me understand all of the potential consequences that might come up in engineering, from both a financial and legal point of view.

I would like to express my gratitude to Mark Griffieon, Joe Montairo, Tony Paolucci, Stephen Engel, Claire Rauser, Todd Clemens, Kevin Larkin, Mark Vanderbecken, Manolo Hoyos, Yordan Yordanov, and a long list of many others I have worked with. Their constant striving for perfection has made me a better roll tooling and equipment design engineer. I'm thankful to them for trusting me to be a part of their projects.

I thank Raymond Aaron for helping me to realise my idea of writing this book.

And finally, I thank my parents for inspiring me to be the best I can be, and my father who guided me into the world of engineering.

Foreword

I am excited to introduce to you Dako Kolev, author of *Rollforming 101: How to Choose the Right Technology*. In this book, Dako presents the basics of rollforming technology, and the associated equipment and roll tooling design principles that you, as a beginner in this industry, should know.

Dako also shares his knowledge and practical expertise in providing rollforming solutions.

Whether you are a rollforming operator, or a rollforming and roll tooling equipment designer/engineer, this book is for you. If you are in a management level and would like to implement the technology of rollforming in your manufacturing operations, this book will help you to foresee the benefits of high speed production.

I urge you to learn how beneficial rollforming can be for your operations by reading this book!

—Raymond Aaron
NY Times Bestselling Author

Introduction

Henry Court (1740-1800), an English inventor, introduced the technology of rolling using rolls by forming red-hot iron into bars, in 1783. His first rolling mill introduced grooved rollers to produce iron bars more quickly and economically than the old methods of hammering. His invention revolutionised the British iron industry.

The idea of forming by rolls has been developed further and applied for cold forming as well. Today's modern cold roll mills are used to bend and form sheet metal through several rolling dies stands in great quantity and speed. Complex metal-formed parts and cost effective production in big volume is what established the rollforming industry and took a significant market share in metal forming.

The technology of roll forming is capable to produce parts with variable complex shapes including elements of holes, edge notching, cut ends additional forming, in line welding, part slitting, swiping, curving, etc. The parts are formed in desired shape, within tolerances, smooth finish surface, cut to length, and ready to install. A configuration of a roll forming line with specific features added to a roll formed part requires thorough and advanced knowledge in designing the process, designing and properly sizing the equipment, designing roll tooling (the art of roll forming), and designing dies and specific tools for specific parts.

The idea of this book is to layout the basics and to explore what the parameters are that drive the complexity of the rollforming technology

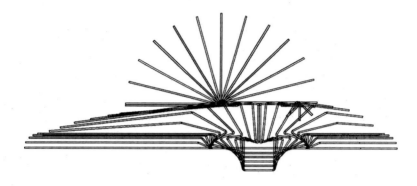

Dako Kolev P.Eng.

CHAPTER 1

Roll Forming Technology Basics

1.1 Roll Forming Definition

Roll Forming is a progressive motion process of forming flat strips of cold sheet metal through several stages of forming or bending, thru pairs of rolls, gradually without changing the thickness.

1.2 Roll Forming Technology justification – top 10 advantages.

If you are an entrepreneur or a CEO, launching a new metal formed product, and your intention is to enter the Roll Forming industry, your first question will be: what is my ROI [return on investment]?

The technology of Roll Forming offers great advantages versus press brake forming.

No 1. *Volume*

An investment in a new roll forming line is justifiable when you have great volume of parts estimated in linear feet (i.e. 1,000,000 feet per year) or in metal [steel, aluminum] weight. Most roll forming companies are estimating their profit as a % of the weight for the total purchase order. If you estimate your volume based on weight, the speed of the roll forming line is essential. Heavy gauge rollforming lines, at high speed, cost significantly more, but if the total metal weight rollformed for a certain period of time is what pays the bills and keeps the profits high, it is an efficient ROI.

No 2. *Ability to produce variety of shapes in the same Roll Forming machine [rollforming line]*

By changing the roll tooling in the rollforming machine, changing the cut-off Die (or just the cut-off shear blades), you can produce parts with a different shape. The cost of the new roll tooling and Dies is added to your selling price, unless your customer would like to be the owner of the tooling and pay for it.

No 3. *Ability to produce parts of similar shape with the same Roll Tooling*

- Different part web
- Different legs
- Specific part members

We design combination split rolls and by changing certain rolls we change certain segments features of the roll formed part. Another technique is using quick change spacers to re-arrange the rolls and have different dimensions of the segments. If we have to design a combination set of roll tooling for similar shape parts, a separate set of shear blades for the cut-off system will be required for each size.

No 4. *Ability to produce parts at variable length*

Cut-off systems are part of the roll forming lines. In a post-cut line configuration with a flying shear or cut-off die, the cut length is pre-set automatically. The electrical controls are designed and programmed to perform different jobs defined by the part's quantity and length. In Europe and Asia, still, many post cut rollforming lines are designed and build type *Stop and Go*. The cut off die is stationary and the Rollformer works on start and stop mode. Speed limits in *stop and go* lines are 45mpm max., for 10ft length. And both types depend on an encoder to activate the cut-off process and the part to be cut at specific length. Length tolerances for different industry are different.

No 5. *Smooth surface on the finished product*

Once the roll tooling is designed and manufactured correctly, the finished surface of the product is smooth without scratches or wrinkled sections. Any marks at the corners must be acceptable if the application has no specific requirements. The roll tooling material we use is mostly D2 and performs perfectly without leaving unacceptable marks on the material. In some special cases forming pre-paint material, the rolls are flashed, chromed, and polished. To avoid any surface marks, additional technology of laminating the metal strip before roll forming is used, covering with a thin plastic layer on both sides, i.e. stainless steel parts for the automotive industry.

No 6. *Forming parts with different thickness – 0.005"–0.500"*

The material thickness range in the rollforming industry is between 0.005" to 1/2". In rare cases, material can go 3/4" or above. When we configure and size the equipment, we size it for certain material thickness range. The advantage is that with the same set of rolls, parts with a specific material thickness range can be manufactured. But the thickness range within is not advisable to be more than 2 times.

No 7. *Integration of additional operations in line*

Another great advantage of the rollforming technology is adding features to the parts being formed. Within the rollforming line, we add specific tooling and associated equipment to do:

- Notching of edges
- Piercing of holes
- Embossing
- Knurling
- Lansing
- Welding – spot, laser, high frequency welding, etc.

No 8. *Post-secondary operations off line*

After the parts are roll formed and cut to length, they can proceed further to post-secondary operations for additional press forming and specific piercing. One of the first post-secondary machines I designed was for wire mesh decks "reinforcing channels" that required **3 post**-secondary operations: spread the ends, bend ends to L shape, and pierce mounting hole. These machines well served the wire mesh decks manufacturers.

No 9. *Rollforming as a part of a complex automated technology with special purpose machinery*

For example, manufacturing steel doors requires rollforming the longitudinal edges of the steel panel and is one of the fourteen line operations. This complex assembly door line (Fig.1.1) from start to finish has bending operation, which for line speed purposes instead of press brake bending is done by a Rollformer between the other special purpose automated machinery.

Rollforming 101

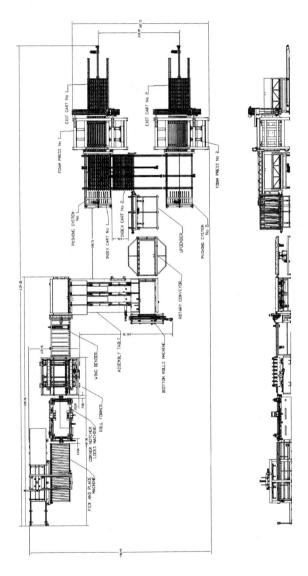

Fig. 1.2 Residential metal door line general layout

No 10. Sufficient cost reduction

Cost reduction depends on volume and volume depends on rollforming line speed. High speed rollforming lines are more expensive, but ROI is faster and with great benefits in a long run. Compared with press brake forming, the cost easily is reduced with minimum 15% to 25%. Average rollforming speed range is 100–250ft/min. High speed can reach 300–450ft/min., i.e. stud/tracks roll forming machines. When you are calculating the projected cost reduction based on volume, you must consider that the efficient roll forming time is 60% only in an 8 hour shift. Still speed is in relation with the part complexity, sensitivity, and heat anticipation, and certain rollformed parts have speed limit established through the tryout process.

1.3 Designing parts for the technology of Roll Forming

When designing parts for rollforming, you must be aware of:

1.3.1 Cross section depth

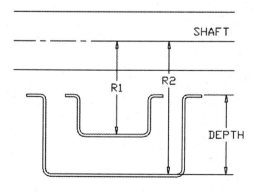

Fig.1.3.1

- Part depth determines Roll's Diameter – Fig.1.3.1 The deeper the cross section, the bigger the Roll's diameter, and controlling the shape of the part is more difficult. In a case of roll forming a part with deep cross section it's recommended to use rollforming machine with ratio bottom to top shafts – 2:1; 1.5:1; 1.33 or any custom ratio. It will reduce bottom rolls diameters and size of the Roll former.

- Deeper cross section will require more passes for smooth run. It is even more complicated if the part is not symmetrical.

- Deeper cross section will require sufficient distance – vertical (bottom to top shaft) and horizontal (pass to pass)

- The cost for the Roll Tooling is higher (if it can fit to an existing Roll Mill. Otherwise a new Roll Mill will be required with a specific custom configuration to meet the part's design requirements).

1.3.2 Web length or flat area

When the part is designed for rollforming technology, the following should be considered:

- ***Fig.1.3.2a*** Length of the flat area. Wide flat cross sections will cause wrinkled surface, and oil canning.

Fig.1.3.2a

- ***Fig.1.3.2b*** If having a wide flat cross section is absolutely necessary – consider flattening Rolls (i.e. after roll forming, the part is additionally bent and cut through a progressive Die in 2 or 4 pieces.) If the panel is not formed in a duplex Rollformer but conventional, consider using support rolls in the middle.

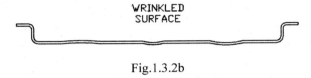

Fig.1.3.2b

- ***Fig.1.3.2c*** If the desired wide flat area is not of importance for assembly purposes – consider adding grooves (arcs, triangle or rectangular shape) for stiffening. Too many grooves as well may cause cross bowing. One or two grooves are desired in the middle of the web.

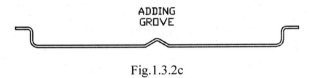

Fig.1.3.2c

1.3.3 Cross section and elements

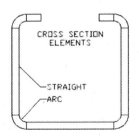

Fig.1.3.3

Every segment of a formed part ***Fig.1.3.3*** consists of 2 elements – straight and curved. When rollforming, we control the formation of the arc (arc bending process) and straightness of the straight elements. Giving certain angle of bending to the arc, the straight element attached to the arc moves and it is important to look at the edge for excessive stretching.

10

1.3.4 Piercing, notching, holes pattern

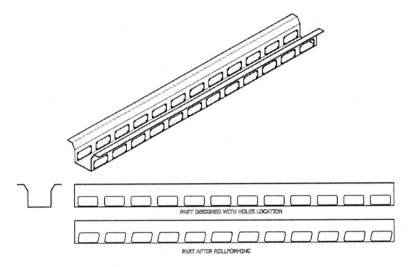

Fig.1.3.4

In rollforming, strain is the worst enemy, and having openings in the material often creates unpredictable side effects. In product design (see *Fig.1.3.4*), *it* is mandatory to consider the following when holes piercing or edge notching is required:

- Holes must be at least 3–4 times the material thickness away from the line of bending.

- Holes must be at sufficient distance from the cut edge.

- Holes close to the part edges will cause wavy edges.

- Round holes at the line of bending will be distorted into an oval.

11

- Having edge notching is considered forming a precut material. The continuity of the strip strength is disturbed and more rollforming stations are required – up to 1/3 more.

1.3.5. *Minimum Radius of bending*

Bending Radius

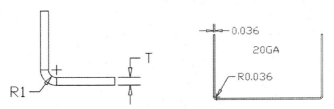

Fig.1.3.5.1 a-b

<u>Common radius</u> – *Fig.1.3.5.1 a-b* – minimum bending radii for forming are provided by the metal supplier or are specified in standards

- The usual minimum bending radius for mild steel is equal to the material thickness.

- In certain conditions it can be 0 degrees

- For high strength steel, low elongation material, the minimum radii must be 4–8 times material thickness.

Large radius

Fig.1.3.5.2

Rollforming is based on the principle that the metal attains permanent deformation at the bend lines – outside fibers strained and inside fibers compressed beyond yield limit. If the metal does not stress beyond the yield limit, it will spring back into flat position.

Forming large radii (*Fig.1.3.5.2*), the tension and compression in the outside fiber can be so small that permanent deformation is not reached. That's why it is very difficult to rollform metals to reasonable tolerances if the bending radius is 10–100 times or more the material thickness.

Small radius

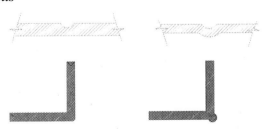

Fig.1.3.5.3

- Very small radius or even 0 degrees can be achieved by hemming, or by special grooving

13

- Attention is to be paid to strip calculation – add extra material

- At bending line there will be reduced material thickness (expect cracking)

1.3.6 Bend Radius – methods of forming

Bend type: Equal

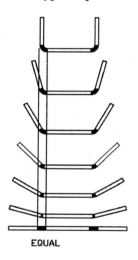

Bends are formed equally from the straight segments on each side of the bend. The length of both straight segments is reduced with each forming station.

EQUAL

Fig. 1.3.6.1

Bend type: Inside

Method of inside bending means that we start forming the ark from outside of the bending segment towards inside and with every pass we increase that ark segment to a full ark.

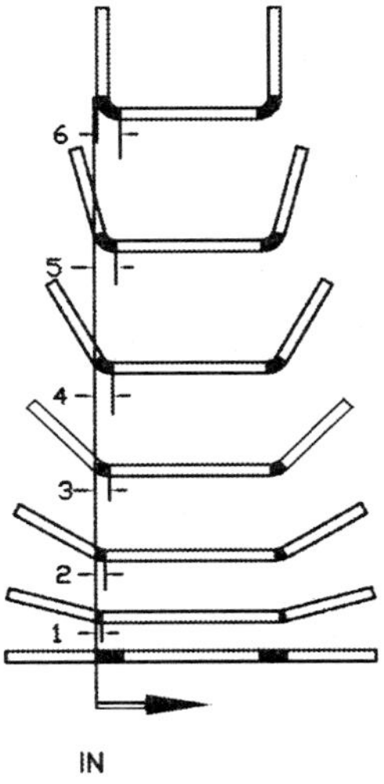

IN

Fig. 1.3.6.2

Bend type: Outside Bend type: Inside

Method of outside bending means that we start forming the ark from inside of the bending segment towards outside and with every pass we increase that ark segment length to a full ark length.

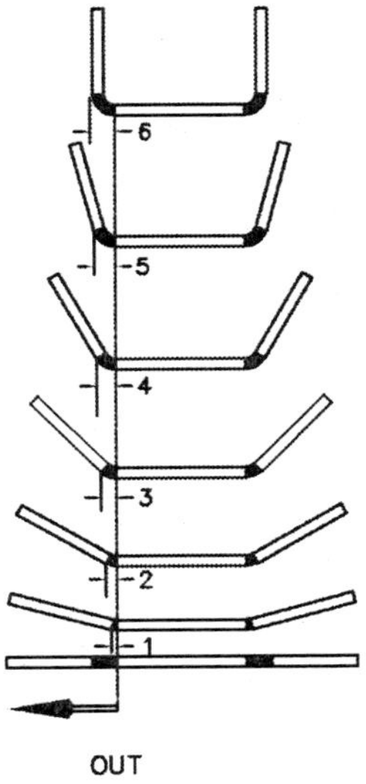

OUT

Fig. 1.3.6.3

Bend Type: Constant arc

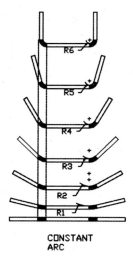

Bends are formed by maintaining a constant bend arc length in every forming station. No change in length of either straight segment is allowed. Only the inside radius is reduced

CONSTANT ARC

Fig. 1.3.6.4

1.3.7 Air bending

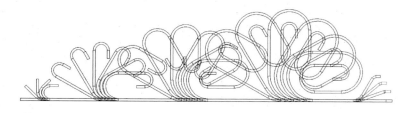

AIR BENDING

Fig. 1.3.7

Rolling the material without supporting the bending radii is called *air bending*. This is an extreme sample (*Fig. 1.3.7*) of air bending where controlling the inside radius after the first round of rolling is impossible. In most cases, we are using side rolls, corner rolls, sliding blocks, etc. to minimize the air bending and keep the forming under control.

1.3.8 Symmetrical and Non Symmetrical Cross section

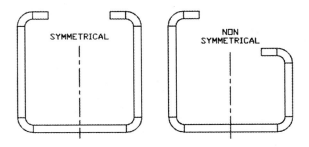

Fig. 1.3.8

Symmetrical parts are not expected to show any side effects if the pressure in the corners is equally set up.

Non symmetrical parts tend to show side effects of twist, bow, or camber. The roll forming energy is not equally distributed on the faces being elevated and the side where the energy and forces of distributed load are bigger will create momentum from the mass center and will twist the part. One of the solutions is to rotate the part on the side of the longer leg to reduce the edge stretching and balance the forming forces on both sides of the part. A well experienced rollforming operator can apply more pressure on the specified corners at the last few stations to minimize and correct the potential twist. Additional exit rotary straightener is used to compensate the twist and straighten the part. Another solution is at roll tooling design stage, the part flower to be developed on angle and reduce the elevation angle of higher leg.

17

1.3.9 Shape of parts for rollforming

We have two major categories – see *Fig. 1.3.9*:

A. Open

B. Closed

Additionally between them they are divided on:

For the category of <u>open</u> profiles:

- <u>Simple</u> – with one to four bending corners.

- <u>Complex</u> – with multiple bending corners and hemmed edges, combination between open and closed.

- <u>Panels</u> – wide profiles with corrugated trapezoid or wave shapes, locking ends, door panels, electrical wide trays, special.

For the category of <u>closed</u> profiles:

- <u>Interlocking</u> – square, rectangular, polygons, round or special shapes

- <u>Tubes</u> – round, square, rectangular, welded seamless tubing

OPEN

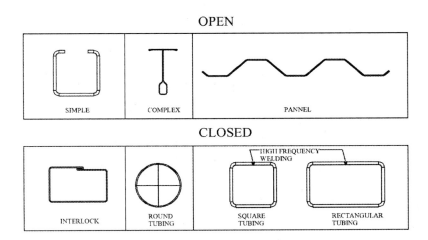

Fig. 1.3.9

1.3.10   Dimensioning parts for rollforming.

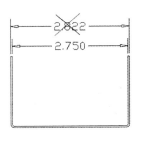

Fig. 1.3.10.1

When design of a part is completed, it is important for the purpose of roll tooling design to establish GDT – general dimensions and tolerances. For assembly purposes, the end user knows

- which dimensions are important and have to be held within the tolerances

- in which direction the material thickness "grow"

(when minimum and maximum material thickness is established)

Dako KOLEV P.Eng.

Fig. 1.3.10.2

For example: If the web dimension is for track application, must be shown inside of the part – *Fig. 1.3.10.1*. Material thickness increase can only go outside.

And if the web dimension is for stud – *Fig. 1.3.10.2* – it is essential to be shown outside of the part. Material thickness increase can only go inside.

1.3.11 Tolerances

- For the purpose of high volume of production (high speed), acceptable tolerances are required in order to justify the technology of rollforming. Depending on the application, the rollformed products can have:

  ○ Loose - +/- .030"-.040"

  ○ Medium - +/- .015"-.025" (average in the industry)

  ○ Tight - +/- .005" (i.e. parts for automotive industry)

Length tolerances

Length tolerances are important for choosing the right die accelerating system.

If the tolerances are within 0.015", the length measuring system must be "close loop." The design of this type of cut-off system is required to implement servo motor where the die acceleration is with timing belt, ball screw, or rack and pinion drive components.

- If the length tolerances are "loose" – open loop measuring system will do the job. The Die accelerating systems in this case are designed with air accelerating cylinder or with spring die return. Spring die return is the least expensive solution and works in slow speed rollforming, where the cut-off knife penetrates into the part and drags the die till it releases it when knife strokes out.

- Positive stop and special features for cut-off die dragging. In slow speed where more precision cut to length is required, there is a solution with positive stop or picking mechanism. The picking mechanism is part of the cut-off die, grabs the part in specially designed and pre-punched hole(s), and drags with the part till cut is completed.

Surface appearance tolerances

- Hot rolled and even cold rolled steel is less sensitive to surface scratches. Still, traces of marks in the corner areas can be detected easily; the question is: are they acceptable for the final product appearance or does the process of forming need to be revised and reduce such marks?

- Pre-painted metal, high luster stainless steel, or aluminum has to be carefully formed to avoid scratches. It is a good practise to consider more passes, different tool steel and/ or surface finish. Flush chroming on the rolls is well proven to work with pre-painted material. D2 is most preferred tool steel.

- Rollforming 302/304 stainless steel for outdoor application. I had the experience of forming panel

302 SS material with D2 rolls in the field. Even after forming, the part's corners were brushed and acid cleaned, the carbon from D2 left over the smashed stainless steel layer, 6 months later start showing signs of rust. Then we decide to change the rolls with stainless steel rolls, special grade, and the rust effect was resolved.

1.3.12 Min/Max Material thickness relationship

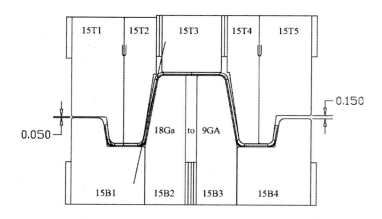

Fig. 1.3.12

Within one set of roll tooling, we can manufacture the same part with different material thickness. By bringing down the top rolls, we set up the rolls for thinner material. What is important to understand is that the material wraps around the roll's corners – Fig. 1.3.12, but the vertical gap for maximum material thickness is not fulfilled – thinner material goes diagonal. If we need to have total control, the rolls must be split and with quick change spacers (shims); the gaps have to be closed.

1.3.13 Material mechanical properties

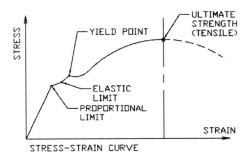

Fig. 1.3.13

When any material is tested under load, a stress-strain curve, fig 1.3.13, presents the approximate limits of yield and tensile strength. When bending, the material must be stressed **beyond the yield point**, to establish permanent deformation and reduce the spring back effect. Permanent deformation of the material is happening after the yield point. **Ultimate strength** or so-called **Tensile strength** is the maximum stress value obtained on a stress-strain curve. Beyond that point is the zone of distortion and break. Tensile strength is considered for calculating the shearing forces in the cut-off systems.

1.3.14 Theory of Bending

Bending and changing the radius of neutral axis

Forming a flat strip of cold metal is a process that is a progression motion through several stages of forming or bending. In any cross section, you can define straight and curved elements. It is assumed that the straight elements do not change their length.

For the curved elements, the neutral axis is changing position and the location depends on

- material thickness

- radius of bending

- material's mechanical properties

Let's look at a curved element and isolate a single fiber or visualize it as a rod (see Fig No1.3.14.1) between sections: 1-1 and 2-2 in an angle dψ.

O1-O2 fiber axe before bending, C1-C2 after bending – neutral axe

The normal stresses (σ)on both sections 1-1 and 2-2 are creating a pair and the line between the section 2-2 and actual line of action will change with ($\delta d\psi$)

To define the normal stresses (σ) between points A1 and A2 in a distance Z from the neutral axe will accept the positive Z direction (of the coordinate system) as an extension of section 2-2.

The fiber A1A2 will elongate A2D2 and the respective stress is:

$$\sigma = \varepsilon E$$

Where ε - the relative fiber elongation A1A2.

$$\varepsilon = A1D2/ A1A2$$

$$\rho - \text{radius (A1A2 fiber)}$$

Therefore: $\quad$ A2D2 $= z\,\delta\,d\psi$

$\qquad\qquad$ A1A2 $= \rho\,d\psi$

$\qquad\qquad$ $\varepsilon = z\,\delta\,d\psi/\rho\,d\psi$

$\qquad\qquad$ $\sigma = (z\,\delta\,d\psi/\rho\,d\psi)E$ $\qquad\qquad$ (1)

Formula (1) based on Huck's Law gives the normal stress distribution in height in relation of bending moment.

As ($\delta\,d\psi/d\psi$) and E are constant, the normal stress σ is changing and depends on variables:

$$z \text{ and } \rho$$

In straight beams, the stress has linear distribution.

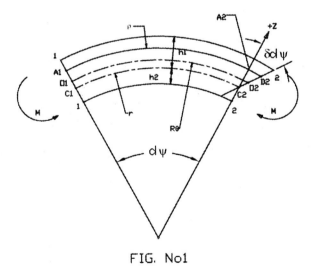

FIG. No1

Fig. 1.3.14.1

In curves, the stress has hyperbolic distribution (Fig.1.3.14.2). It can be seen that the stress in positive direction (+) Z is increasing slower than those in (-) Z direction.

Thus, in a curve section, the stress is bigger in inner side than in outer side of the element.

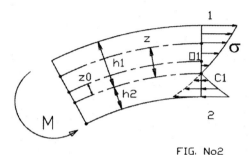

FIG. No2

Fig. 1.3.14.2

If we solve the static equation:

$$\sum X = 0; \quad \int_F \sigma \, dF = 0; \qquad (2)$$

And

$$\sum My = 0; \quad M - \int_F \sigma \, dFz = 0; \quad (3)$$

$$\int_F \acute{o} \, dF = \int_F E\left(z \, \delta \, d\psi \, / \, \tilde{n} \, d\psi\right) dF = 0$$

As we solve it:

$$\int_F \left(z \, / \, \rho\right) dF = 0 \qquad (4)$$

26

This equation allows us to find the position of the neutral axe. As we exchange

$$z = \rho - r \qquad \text{(Fig. No1)}$$

$$r = F / \int_{F} dF / \rho = 0$$

1.3.15 Bending allowance

Bending allowance defines the length of a straight segment required for each bend.

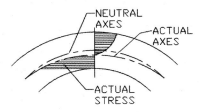

Fig. 1.3.15.1

Fig. 1.3.15.1 shows the bending and stress distribution in the process of rollforming and the minimum radius of neutral axes.

In reality, the offset of the neutral axes and the radii is as shown in the figure 1.3.15.2 and is identified as "K" factor – percentage of the material thickness.

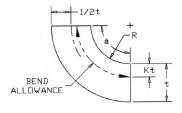

Fig. 1.3.15.2

Dako KOLEV P.Eng.

All that refers to roll's tool designers, who need to foresee the behavior of the material in each stage of forming and properly design the tooling. There is roll tooling design software in the market with sophisticated features as FEA and simulations that help the roll tooling designer to foresee and predict any side effects. Still there are many unknowns in this area and only the designer's expertise is the key in designing roll tooling.

Recommended bend radii can be found in standard steel specifications (ASTM, DIN) or by the steel supplier.

As a rule of thumb:

- For materials 30-60kpsi yield bending allowance is 28–33% inside radius

- For stronger materials – 80kpsi yield and higher, it should be considered close to 45–50%

1.3.16 Spring back and overbending

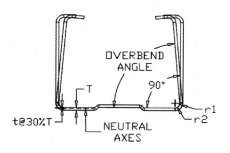

Fig. 1.3.16

When bending, not all layers of the material perform plastic deformation. There are layers that are still in the elasticity zone and tend recovery. That causes the material to spring back after bending force is released.

The bending spring back is a function of: material thickness, inside radius, k-factor, module of elasticity, and yield strength (psi).

In rollforming to compensate for spring back in material thickness 20 to 12 Gauge recommended overbend would be 3-5° and in range of light gauge 24-20Ga – 5-7° or 7-9° (see Fig 1.3.16). If you search the internet, you will find few spring back calculators with proposed formulas for calculating the expected final angle. I it is common practice to design overbend rolls with 2 different overbend degrees on each side of the roll, and in the try out process we decide which one works the best for a specific part. Additionally, corner rolls fixtures with fine tune adjustment features can be added to the rollforming process if necessary.

1.4 Side effects in rollforming reflecting to straightness and flatness

1.4.1 Side effects (Fig. 1.4.1 and Fig 1.4.2) are expected in Rollforming such as:

- Longitudinal bow – When pressure of the top rolls applied exceeds in certain station (pass). Recommended – hump pass, before the last pass

- Camber (sweep) – when uneven pressure is applied from the rolls from left side versus right side in the same rollforming stations.

- Twist – in non-symmetrical parts, the side with more corners to roll form will absorb more rollforming energy and will twist the part in the same direction.

- Flare – edges at cut line will be opened outside of cut-off Die and closed inside.

- Cross bow – the bottom (web) of the part will deviate from straight to round

- Wavy edges – excessive stretching of edges. It can be seen in cases of extreme and aggressive edges elevation within a short distance. Rollformer's horizontal distance (pass to pass) is too small.

- Oil canning – is a moderate buckling of sheet metal in the flat area of a rollformed part. It is seen as waviness and is common in light gauge material, where excessive pressure in the corners is stretching the material.

- Chevrons – V- shape wrinkles in the corners, seen in light gauge and ultrathin material. Prior to from the corner, the flat area already is having oil canning effect.

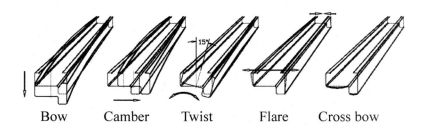

| Bow | Camber | Twist | Flare | Cross bow |

Fig.1.4.1

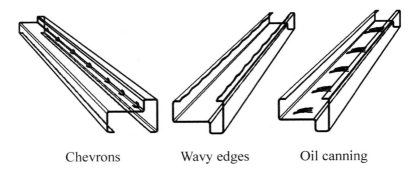

Chevrons   Wavy edges   Oil canning

Fig.1.4.2

1.4.2 Contributing factors for the above side effects:

- Non-symmetrical parts

- Tight tolerances

- Not appropriate bending radiuses with a respect of material thickness and material properties

- Not sufficient roll tool design and incorrect part's flower development

- Holes are too closed to the edges in pre-notching and pre-punching stage

- Residual stress concentration in the material

- The Rollformer is too old and needs complete refurbishment. Bearing housings and take-up bearing blocks are out of manufacturer's tolerances, increased clearance, radial and axial movement is in place. Often is seen how the top shaft is moving up and down when the part is rollformed – elevation nut assembly issues

- The Rollformer's shafts are not aligned prior to loading new set of rolls

- The roll tooling is not set up properly

CHAPTER 2

Roll Forming Equipment, Design Features and Applications

List Of Equipment For Rollforming Line Configuration

Uncoilers

- Coil cars
- Seizer or treaded tables
- Flatteners/Levelers
- Feeders – Servo
- Flatteners – Feeders combination
- Pre-punching presses – hydraulic , mechanical, air
- Punching, pre-notching tooling – Dies
- Roll Forming Machines
- Cut-Off Systems
- Packaging systems

- Additional inline operations – Curving, welding, embossing, lansing, printing etc.

- Special purpose machinery for post-secondary operations

A rollforming line configuration is consisting of 3 systems:

Entry systems
Rollformers
Exit systems

2.1 Entry systems

Prior to rollforming, the material is processed thru the following machineries:

2.1.1 Uncoilers

- Basic technical specification for Uncoilers:

- Mandrel type – single, double, turret

- Mandrel capacity – 2,500lb, 5,000lb, 8,000lb, 10,000lb, 15,000lb, 20,000lb

- Mandrel expansion – min/max, manual, or hydraulic

- Mandrel Face – 12" to 50" (coil width)

- Mandrel leaves type – 3 or 4, tapered blocks or linkages

- Brake type – Electrical, Air

- Mandrel drive:

- Driven – feed up drive, constant drive

- Non driven – pull thru

- Base adjustment: a/ fixed, b/ manual, c/ hydraulic

- Loop control – dancing arm with limit switch, laser, ultrasonic sensor etc.

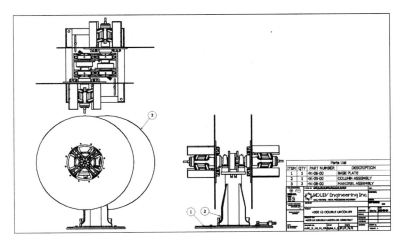

Fig.2.1.1

2.1.2 Coil cars

Coil cars are supplementary machinery that works in combination with Uncoilers for coil handling. Featured by:

- Max capacity

- Max coil OD

- Max coil width

- Cradle width

- Cradle lift – 24"

- Means of lift (hydraulic cylinder)

- Coil car drive – hydraulic motor

Flattener (Straightener)

Purpose: To remove coil set in material. If the part requires pre-punching and/or pre-notching, this machinery pulls the material from the Uncoiler and runs it thru a set of staggered rolls (Fig. 2.1.3a – ZMM Pobeda Ltd.).

Features:

- Process material – 12" to 72"width, thickness – form .010" to .500"

Fig. 2.1.3a

Rollforming 101

- Straightening Rolls – 5, 7, 11, (Optional + 1 for fine tune up), size: 2" –6"

- Back-up Rolls for deflection prevention available

- Cluster gear drive

- Top rolls independent adjustment

- Entry guide system

- Entry pinch rolls

- Exit pinch rolls (Optional – adapted as servo feed rolls)

- Pinch control – air cylinders (hydraulic for heavy gauges)

- Exit basket rolls

- Heavy duty stands and base

- AC, DC variable speed or Servomotor drive

- Operators console

- Jog remote control

- Optional Custom Solutions

Percent penetration

Flattening (Straightening) is accomplished by bending the strip around sets of rollers to stretch and compress both surfaces, exceeding its yield point. The result is equal surface length after spring back or flat material.

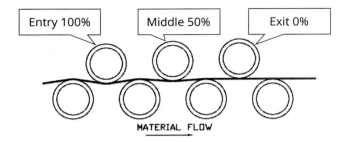

Fig. 2.1.3.b

Standard percent penetration recommended is shown on Fig. 2.1.3b.

2.1.4 Feeders – Servo

Purpose: to feed the material into a pre-punching system in precise length progression.

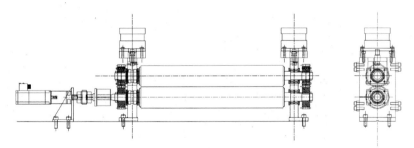

Fig. 2.1.4.1

Rollforming 101

Features:

Fig. 2.1.4.2

- Rolls diameter range from 2 ½" to 6"

- Rolls hardened surface, grind and sand blasted, chromed, satin finished

- Roll space: to accommodate material width 12" to 72"

- Material thickness – up to 3/8"

- Pinch control: air or hydraulic cylinders with pressure regulators

- Back lash control: split gears set

- Cabinet base if stand alone or mounting plate to the press

- Entry basket rolls

- Entry guiding system

- Drive: servomotor – closed loop

- Variable feed settings

2.1.5 Pre-punch presses

Presses (Fig. 1.5) used for pre-punching operations in rollforming are hydraulic, pneumatic, and mechanical. For accurate feed progression, the presses are working with attached precision servo feeders (Fig. 1.4.2) synchronized with the press stroke.

High speed rollforming lines will require 50–100 strokes per minute at the entry pre-punch operation. Accumulation pit between the pre-punch press and Rollformer is a common feature in this type of line configuration.

Fig. 2.1.5

2.1.6 Shears in pre-cut line configuration

In <u>Pre-cut rollforming line configuration</u>, where the sheet is pre-cut to length prior to enter the Rollformer, shears are used to cut to length – Fig. 2.1.6.

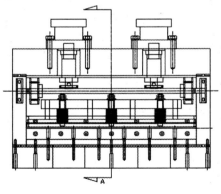
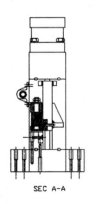

SEC A-A

Fig. 2.1.6

2.2 Roll forming machines – basic types and applications

Major technical specifications for every rollforming machine:

- D – Shaft Diameter (and keyway)

- RS – Roll tooling space

- HD – Horizontal distance

- VD – Vertical distance – min

- VA – Vertical adjustment – max

- BD – Base to bottom shaft

- Ratio bottom to top shaft

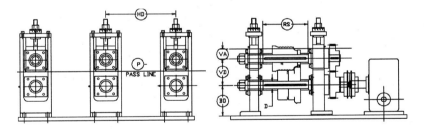

If a set of rolls is to be designed and load for a specific rollforming machine, the above technical specifications are necessary.

2.2.1 Dedicated

Applications: These types of Rollformer are designed and built for dedicated product/products. All technical specifications, Rollformer size, ratio, horse power, and speed are specifically adjusted to a specific product.

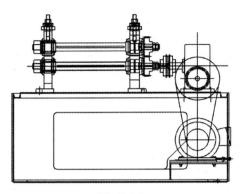

Fig.2.2.1

Features – Fig.2.2.1:

- Inboard stands with bottom to top gears driven
- Bottom to Top shaft fixed ratio
- All Bottom shaft driven by worm gear reducers

2.2.2 Cluster gears Rollformer

All bottom and top shafts are geared with ability for vertical adjustment

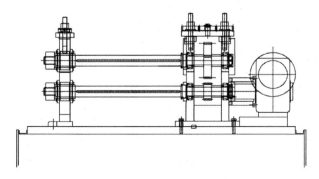

Fig. 2.2.2

Features – Fig. 2.2.2:

- Inboard Continuous Gear Reducer
- Fixed ratio
- Limited Vertical adjustment

Applications: It is a less expensive version of dedicated Rollformer. Build for a family of dedicated product/products with fixed ratio. Any new product and set of rolls loaded on this Rollformer must be within the horse power available and maximum speed.

2.2.3 Cantilevered Shafts Rollformer

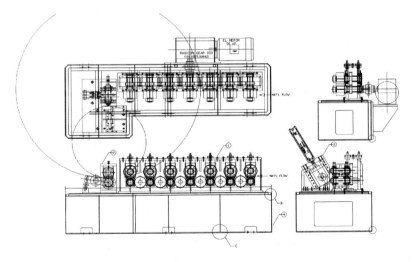

Fig. 2.2.3

Features – Fig. 2.2.3:

- Continuous Cluster Gear box
- Short roll tooling space – up to 4"

- Quick Roll tooling change
- Easy service and alignment

Applications: When Roll tooling change is required often: access, alignment, and roll tooling set up is easy and quick.

2.2.4 Cassette type Rollformer

Applications: Cassette type roll mills are developed for quick change production from one rollformed part to another. The roll tooling for certain profile is loaded and set for each cassette. With quick disconnect couplings, changing cassettes is fast.

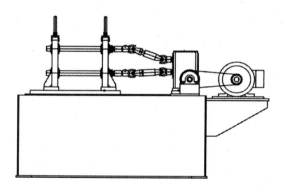

Fig. 2.2.4

Features – Fig. 2.2.4:

- Inboard Stands with quick disconnect couplings or universal joints

- Outboard stands – standard conventional Rollformer design

- Drive – bottom shafts only or both: bottom and top

- Shaft Diameters range: 1 ½", 2", 2 ½", 3", 3 ½", 4". For ¼" material and up, shafts can be 5" and 6"

- Cassette – one cassette consists of minimum 4 rollforming stations (standard)

- Worm gear reducer for each station – train or split drive

- Heavy-duty base with location pins or grind corner blocks for cassettes mounting

2.2.5 Gear-head Rollformer

Applications: Gear-head Rollformer is the working horse in the industry. Used for multiple sets of roll tooling at variable part section depth. Options: by changing the linkages and gears inside of the gear head, different gear's ratio can be established.

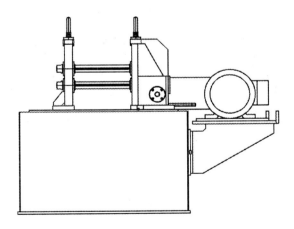

Fig. 2.2.5

Featured – Fig. 2.2.5:

The main feature is that the inboard stand of the rollforming station is a worm gear reducer where both bottom and top shafts are geared in any given top shaft elevation.

- Constant geared bottom to top shafts at variable vertical distance

- Change gears ratio within the gear-head reducer

- Ability to accommodate rolls with wide range of diameters and parts with different cross section depth

- Shaft Diameters range: 1 ½", 2", 2 ½", 3", 3 ½", 4". Rollformed materials can range from .015" to .188" thick. For extra heavy duty applications, shafts can be 4"

2.2.6 Duplex Rollformer

Applications: Duplex Rollformer is developed with automatic adjustability to form parts with different web width. Both sides of the forming parts are usually symmetrical. There are profiles and panels with non-symmetrical sides as well, and work well with this type of Rollformer. Shafts are cantilevered (stub) with double bearing housings. Typical samples: electrical panels and door panels – residential and commercial, commercial appliances panels, studs and trucks with fixed leg length, etc.

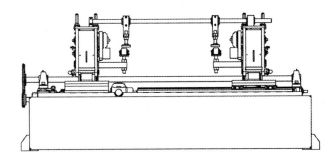

Fig. 2.2.6

They are built in 2 versions – Fig. 2.2.6:

- Centered – both forming heads are adjusting equally according to the center

- Fixed inboard side – the inboard (drive) side is fixed and only the outboard side of the Rollformer is adjustable

- Standard sizes shaft diameters: 1 ½", 2", 2 ½"

- Continuous cluster gears rollforming heads

- High speed of forming – 350fpm for light gauge (or even higher)

- Automatic open/close adjustment of the rollforming heads

2.2.7 Double duplex Rollformer

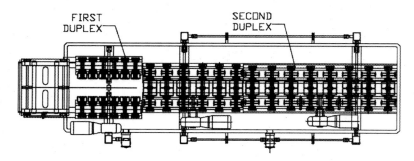

FIRST
DUPLEX

SECOND
DUPLEX

Fig. 2.2.7

Applications: Double duplex Rollformers – Fig. 2.2.7 are special design machinery for "C" channels, studs and panels where the leg and return length is variable. The first duplex forms the return in relation with the leg height. The return size is 1/2", 3/4", or 1" for heavy gauge studs/channels. The second duplex forms the web in relation with the leg height as well. Leg sizes can vary from 1 5/8" to 4", the web for studs up to 14", and electrical trays panels up to 36". Opening and closing of the machine (respectably the web width) is according to the center.

2.2.8 Shaft thru Rollformer

Application – Fig 2.2.8: Another version of duplex Rollformer where the shaft slides through the outboard stands for web width adjustability. Inboard side is fixed and the outboard side is adjustable.

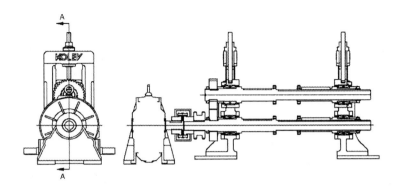

Fig. 2.2.8

Inline additional operation can be added, such as welding, slitting, swiping, curving, printing, Lansing, and embossing. These operations are not a subject of detailed description in this book.

3 EXIT SYSTEMS

Completion of the rollforming process ends with cutting the rollformed part at desired length. Just before cutting, prior to entering the cutoff system, at the exit of the Rollformer, special straightening fixtures are added for final tune up and calibrating the part, fixing bow or twist, and helps entry into the cut-off Die/shear

3.1 Cut-off equipment

In post cut technology, the cut off solutions can be described as follows:

3.1.1 Cut-off presses with flying cut-off die.

The cut-off process can be completed with slug or without. Sample cut-off press with die accelerating system is shown on Fig. 3.1.1. Each cut-off type requires specific cut-off design solutions. Die acceleration system is built in the press. Two type of acceleration are available – open loop and closed loop. If the cut-to length tolerances are not within 0.015", the system is open loop – with air die accelerating cylinder. Average line speed is 200 Fpm. If the cut tolerances are within 0.015", closed loop solutions is used – servo motor with ball screw acceleration. Cut-off presses used are Hydraulic, Air, or mechanical.

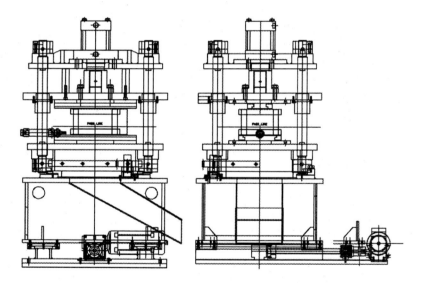

Fig. 3.1.1

50

Flying cut-off air or hydraulic press is a solution where the press with the die is on a fly. Features closed loop acceleration for line speed average 200ft/min. and up to 50strokes/min.

3.1.2 High speed flying shear

This type of flying shears are using slugless shear blades and a powerful short stroke hydraulic cylinder, attached to the sliding shear plates. The design (Fig. 3.1.3) is suited for high speed rollforming operations up to 350–400ft/min. for minimum 8' part length. It is on linear tracks, accelerated with ball screw or timing belts by servo motor – closed loop controls, a well-established combination for speed and cut to length accuracy.

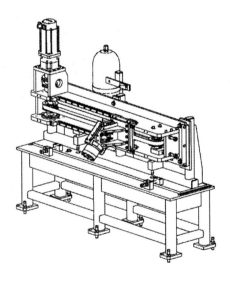

Fig. 3.1.3

3.2 Exit transfer systems

Exit transfer system Fig. 3.2 is a combination of a few machines: exit run out conveyor, parts accelerator, stacker, and transfer system. The system is designed to move the part away from the shear, accelerate it, and create a time for stacking procedure. Once the stacking is completed, the stack is moved away by another transfer system.

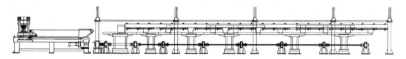

Fig. 3.2 Exit system for roof panels – ZMM Pobeda Ltd.

3.3 Post-secondary special purpose machinery and operations

If specific features of forming and punching are required to be implemented in the part and are in contradiction with the rollforming process, these operations are completed separately as post-secondary operations. For that purpose, special purpose automated machinery are designed and operate as a standalone or as an extension of the rollforming line, continuously by automated transfer systems.

Fig. 3.3

Fig. 3.3 shows additional punching and forming automated machine for wire mesh decks reinforcement channels.

CHAPTER 3

Roll Forming lines configuration

By type of operation, rollforming lines are designed Pre-cut and Post cut.

PRE-CUT OPERATION

In this type of operation (Fig. 3.1), the sheet metal is cut to length prior to entry of the rollforming machine. It is used to roll form parts with complex shape where cut-off would be an issue.

The advantage is that the exit cut-off system is eliminated, there are no cut-off dies to change, or expensive cut-off blades.

The disadvantage is that this type of rollforming line will require Rollformers with 1/3 more rollforming stations and rolls to keep control and pull out of the machine the rollformed part. Trapping and guiding the material from the beginning almost to the end is a necessary rolls design feature that cost extra rolls.

Dako KOLEV P.Eng.

Dako KOLEV P.Eng.

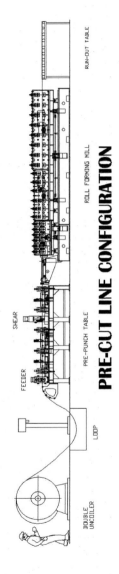

PRE-CUT LINE CONFIGURATION

Fig. 3.1

POST-CUT OPERATION

Most of the rollforming lines are designed with post cut operation – Fig.3.2.

They are designed and configured in 2 modes: Loose and Tight

3.2.1 *Loose*

Typical loose line configuration:

- Uncoiler

- Straightener/flattener

- Strip Loop control

- Pre-punch press with servo feeder and pre-punch/ notching die

- Strip Loop control over a pit accumulator

- Rollforming machine with entry guide, exit straightener, and encoder fixture

- Cut-off Press with Cut-off/shear Die on a fly (Die accelerator)

- Run-out conveyor

In loose line, the material is pulled from the Uncoiler by a driven straightener. Loop control keeps the sheet loose between the straightener and pre-punching/notching station. The pre-punching station is a combination of a press with accommodated punch/ notching die and the material is fed into it by a servo feeder, mounted at the entry side of the press. By creating a loop, the servo feeder will have less resistance when a feeding is in motion with precision progression. Second loop between the pre-punch press and Rollforming machine is needed to establish an adequate timing for the pre-punch press to complete its cycle without the Rollformer, to pull the material and interfere with the pre-punching process. Minimum and maximum loop is established and Rollformer speed is set up within the loop range. After the Rollformer, the part is entering the exit system – cut-off press with flying cut-off die accelerator (or stationary if the post cut line is in mode "STOP and GO") and proceeds further to material handling system.

3.2.2 Tight

The idea of tight line is to speed up the rollforming process and eliminate the expensive servo feeder application. The pre-punch press is built with Die accelerating mechanism and pre-punching operation is completed on a fly. The tolerances of the punching pattern usually are loose.

Additional operations can be added inline, such as: welding, slitting, swiping, curving, printing, Lansing, embossing, stitching, etc.

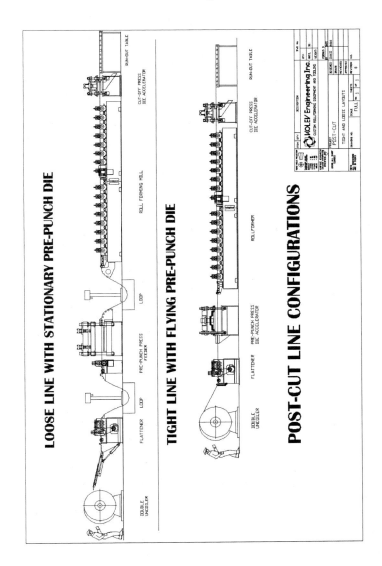

Fig.3.2

CHAPTER 4

Roll Tooling design basics

Designing rolls is an art and takes years to master and develop a high level of skill and vision. Knowledge in using high advanced roll tooling design software is a must, but that has to be backed up with practical knowledge and experience. Good engineering understanding in strength on materials, metals structure, and physics is what a roll tooling designer must know.

4.1 Basic considerations

Prior to roll tooling design and manufacture, the following has to be taken under consideration:

- Quantity of parts that has to be manufactured with one set of rolls (rolls life time)

- What is the material of the product that the rolls have to work with – CRS, HRS, SS, GALV, AL, PAINTED, etc.?

- Tolerance requirements – GD&T (general dimensions and tolerances)

- Roll tooling changeover – consider frequency of changing, cassette type?

- Is the tooling for single purpose or multiple? For example, roll forming "C" channels with different width (Fig. 4.1) – adding/removing spacers (both sides shafts support)

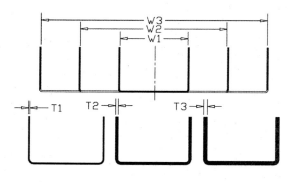

Fig. 4.1

4.2 Roll Tooling Material

Rolls are manufactured with different tool steel and are based on the application, life to last, and price. Most common tool steel used for rolls in North America is:

- 1045
- 4140
- 4340
- 8620
- 52M

- L6
- A2
- O1
- D2

For 1045, 4140, and 4340, movement and deformation after heat treating is a usual effect and that requires rolls to be chromed for good finished surface.

A2, O1, and D2 are having minimum deformation after heat treating and after CNC re-cut, the surface is smoothed and lasts a lot longer.

The most preferable tool steel for rolls with great quality is D2 with minimum to NONE deformation after heat treating and with the final re-cut, the rolls work with all materials used for rollforming parts including pre-paint materials without leaving scratches on the surface. If rollforming Stainless steel materials, and the parts are to be used in outdoor applications, the tool steel's carbon leaves a thin layer on the surface and the stainless steel part after time tend to rust. It is recommended that the areas where the rolls have been in contact with the stainless steel material be cleaned with acid cleaner, or the material prior to rollforming be poly covered.

4.3 Rolls manufacture procedure

The following is a sample procedure for manufacturing rolls. Once the rolls are designed, list of blanks is completed with finished rolls diameters and finished length. The tool steel supplier is supplying rolls, diameters closed, to the finished ones, and cuts each roll with length (+) 1/8" to 1/4" extra for cleanup and finishing grinding.

4.3.1 Rough Bore – this is an operation (many rolls manufacturers are still using) for drilling the roll on a dedicated lathe.

4.3.2 Once the rolls are rough bored, precision CNC machining is completing the following 4 operations:

- CNC Bore machining (-.003")

- CNC one face

- CNC profile machining – plus 0.015" extra material for re-cutting after heat treating

- Flip over and CNC machine the other face and remaining profiling. Extra material for grinding to leave: +0.003" per face

- Machine groove for rolls identification – always on the operator's side.

4.3.3 Punch identifications symbols in the identification groove: Roll # (1T1, 1T2, etc.), Material (D2, O1, etc.), Rolls set ID #, Name of the company that manufactured the rolls.

4.3.4 Keyway broaching – size is usual + 0.015" to 0.020" bigger for easy loading and unloading the rolls. In specific cases, the keyway can be tight – (+) 0.003" to (+) 0.005" and is achieved by EDM key cut.

4.3.5 Heat treating is the next procedure where, depending on the material, it can be hardened through with hardening

58–60 Rc., or just surface flame hardening (for chrome finishing).

4.3.6   Grinding – bore and both faces are ground to size.

4.3.7   CNC re-cut – this is the final operation where the profile of the roll is completed as per profile outline and surface fine appearance. Material to re-cut – 0.015"-0.020" all over the profiled surface.

4.4 SPACERS

Spacers are part of the roll tooling to fulfill the unused space on the shaft. Type of spacers used care – Standard or special.

- ***Standard spacers*** are bushings usually from mechanical tubing, bored to Rollformer's shaft diameter (+ 0.015" to +0.020") and ground to length. In some cases, they are required to be with keyway.

- ***Special spacers*** are types of spacers for roll tooling quick change over. They are available with different design features like a horse shoe with counter bore or split. When the rolls need to be rearranged, the quick change spacers easily can be positioned between the rolls in different locations for different rollformed profile size.

  Material used for special spacers is preferable to be high carbonized, i.e. 4140 and additional heat treated to 50–54Rc.

4.5 Other factors to be considered when roll tooling design

- Part orientation – most parts are preferred to be roll formed with legs up. It is the roll tooling designer's decision how to develop the flower of the part and in which direction the edges (part's legs) are to be positioned and orientated at the exit – straight up, straight down, or in specific angle. It is important in the whole rollforming process that all important bending corners are to be in full contact with the rolls. At the exit, how the part is positioned affects the cut-off operation as well.

- Pass line – is it constant or variable?

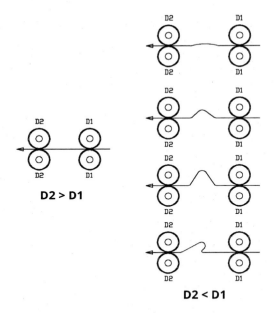

Fig. 4.5

In this Fig. 4.5 are shown 2 cases:

a. When the roll's diameter in every sequent pass is bigger than previous. In rolls design we establish rolls progression within 0.010".

b. When the roll's diameter in every sequent pass is smaller than previous. This is negative progression and the diagram shows the final result of buckling the material between the roll forming stations. In some cases, this technique is used with the idea that the part mass center to be in one line. Stretching at the edges is reduced, but to avoid the above shown undesired effect of buckling, most of the top shafts must not be driven but idle.

4.6 Sequence of forming bends

When developing the part flower, the roll tooling designer can proceed with one of the following methods of forming bends:

- One by one

- By pair

- Few at the same time

- All at the same time

- Start bending from the edges inside

- Start bending from center towards outside

- Alternatively bending

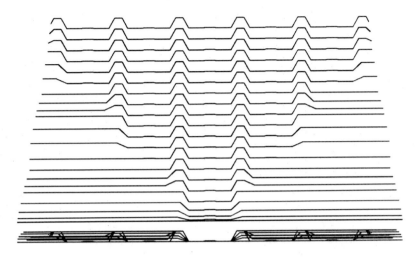

Fig. 4.6.1

Sample of bending Fig. 4.6.1 "From the center towards outside" – this is the most common solution in forming symmetrical parts. All forming bends are done by pairs. It is a smooth process with balanced energy distribution and no side effects.

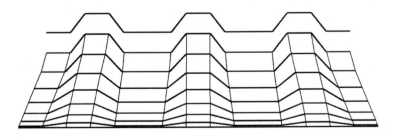

Fig. 4.6.2

Sample of bending – "All at the same time" is shown on Fig. 4.6.2. In this type of forming, all bending corners are engaged at the same time. In each pass, the forming bend location is moving from outside towards inside. The stretching may take an effect in the final shape and flatness appearance.

4.7 Edges and bending corners movement

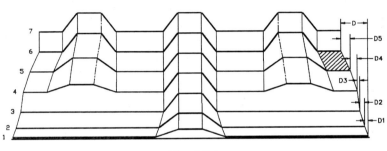

Fig. 4.7.1

The above diagram – Fig. 4.7.1 illustrates how the edges are moving from one rollforming station to the next.

Let's take a look at the section – Fig. 4.7.1 – and describe the stretching:

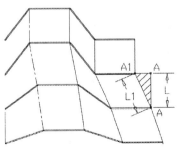

- At edges (or corners), point A is in location of point A1 at the next pass

- Result is stretching – δ = L1/L

Fig. 4.7.2

- If exceeds 10% – expect wavy edges, or variety of side effects "δ" can be reduced by:

- increasing Roll mill horizontal distance

- Decrease bending angle for each pass and add – if edges elevation is too aggressive – a smaller bending angle is preferred.

4.8 Using Side Roll Passes

When additional control in the rollforming process is needed, special fixtures with rolls are designed and positioned between the rollforming stations. Designed with adjustability of the rolls as:

a. To help bending over 90°

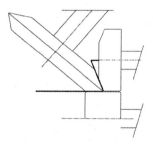

Fig. 4.8.1

The side roll in this case (Fig. 4.8.1) helps by supporting the forming side of the part over 90°

b. Inline side rolls to avoid air bending:

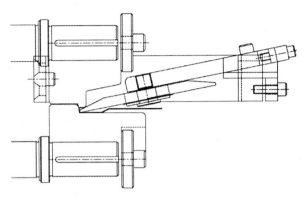

Fig. 4.8.2

Side rolls fixture (Fig. 4.8.2) supporting the lower corner of a "J-bend" panel. Prevents air bending and helps forming inside bend with required inside radius.

c. Standalone fixture between the passes – Fig. 4.8.3

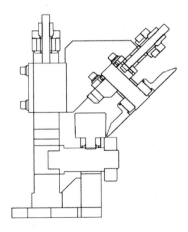

Fig. 4.8.3

d. False bend

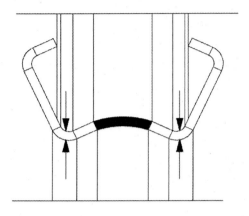

Fig. 4.8.4

Instead of Side Roll Passes in the rollforming process, using this technique for corners control is False Bend – Fig. 4.8.4. It avoids cross bow and round web.

4.9 STRIP WIDTH CALCULATION

All roll tooling design software automatically calculates the strip width for every part designed, based on material thickness and "K" factor – bending allowance.

If it has to be calculated manually – the strip width is the sum of all straight segments lengths and arcs segments lengths. Arcs length

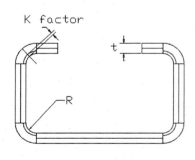

Fig. 4.9

72

is calculated at the location of the neutral axes that is offset and defined by the "K" factor.

K-factor reference info:

For materials yield:

- Up to 30,000psi yield:

  K = 28% (0.28t)

- 35,000psi to 45,000psi;

  K = 30% to 33% (0.30t - 0.33t)

- 45,000psi to 65,000psi:

  K = 33% to 38% (0.33t – 0.38t)

- 65,000psi to 80,000psi

  K = 38% to 42% (0.38t – 0.42t)

- Over 80,000psi:

  K = 42% - 50% (0.42t – 0.50t)

For tryout of a new set of rolls, we start with the calculated strip width. Through the process of the part development, certain part's segments will stretch and the final strip width for full production is confirmed after the try out is complete and all GDT (general dimensions and tolerances) are within the required ones for the part application.

4.10 Define No. of Rollforming Stations (Passes)

There is NO method for precise # of forming passes calculation.

The most advanced roll tooling design software are offered with FEA build in the program and calculate for each forming station the stress of every bend. If the pressure exceeds the assumed internal stress parameters, the designer needs to increase the quantity of forming station and angles elevation, and reduce the stress.

Factors to be considered defining No. of Passes:

1. Cut-Factor:

    • Continuous strip

    • Edge Notched

    • Pre-cut

2. Total degrees of bend – one side from center

3. Number of special passes – grooving, embossing etc.

4. Maximum height of section

5. Material thickness

6. Tolerance factor – loose, medium, or tight

7. Shape factor – simple, medium complex or complex

8. Yield

9. Ultimate tensile strength

4.10.1 *Approximate PASSES estimation*

FORMULA: $\underline{N = (B1 + B2 + B3 + B4 - 0.5)*S + C + G + R}$ (4.10.1)

Where:

- $B1 = $ sq. root($10*H$)
  - where $H = $ section height
- $B2 = 20*T$
  - where $T = $ material thickness
- $B3 = Y*Y/(40000*UTS)$
  - where: $Y - $ yield, UTS $-$ ultimate tensile stress
- $B4 = AT/90$
  - where: $AT = $ angle total degrees of bend one side only
- $S = $ shape factor (simple: $S=1$; medium: $S=1.05 \div 1.1$; complex: $S=1.2 \div 1.3$)
- $C = $ cut factor (Continuous strip: $C=0$, Edge notched: $C=0 \div 2$; Precut: $C=2$)
- $G = $ Number of special passes $-$ grooving, embossing: $G=1,2,\ldots$
- $R = $ Tolerance factor: (Loose: $R=0$; Medium: $R=0 \div 2$; Tight: $R=2$)

4.10.2 Example

Let's estimate the quantity of rollforming stations for "U" channel – 14 Ga, commercial steel, size: web 2" X 1.5" leg – Fig 4.10.2

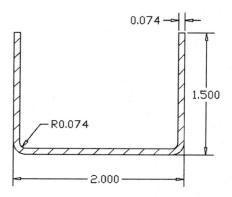

Fig. 4.10.2

- Mat'l thickness: T=0.074"

- Section height: H=1.5"

- Angle to bend: AT=90°

- Cut: for continuous strip C=0

- Number of special passes: G=0

- Tolerance factor – medium: R=1

- Shape factor – simple: S=1

- Yield = 45,000psi

- UTS = 60,000psi

Using the 4.10.1 formula:

$$N = (\text{sq. rt. } (10*1.5) + 20*0.074 + 45000*45000/(40000*60000) +90/90 - 0.5)*1 + 0 +0 + 1$$

$$= (3.87 + 1.48 + 0.84 + 1 - 0.5)*1 +1$$

$$= 7.7$$

Total number of PASSES = 8

4.11 Holes sizes and locations

Holes locations (Fig. 4.11) must be 3 ÷ 4 times material thickness from edges of bending and between them. If less, the holes shape is changing to oval.

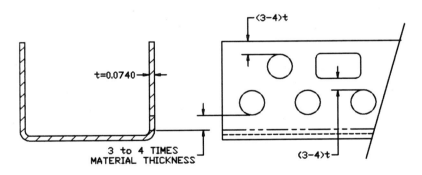

Fig. 4.11

Notching is edges disturbing and the second edge in the run direction is hitting the rolls in every rollforming station. As a result, in every notching pocket, the second edge is bending more than the first edge. Designing rolls for parts with disturbed edges (pre-notched) is considered pre-cut type of operation and requires 1/3 more rollforming stations. Final calibration straightener is recommended to straighten and align the notched edges in the rollformed part.

4.12 FLOWER DEVELOPEMENT

Flower diagram shows how the part changes shape in each step of roll forming – from pass to pass

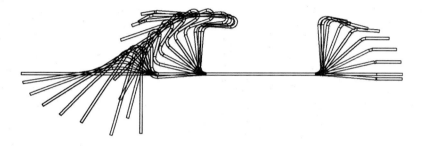

Fig. 4.12

4.12.1 Evaluation Procedure

- Count all curved segment LH and RH on a finished cross section

- Evaluate the quantity of bending steps for each bending segment

- Pass line – is it constant or deviates?

- Part layout for roll forming – legs up? Down? Or in angle?

- Evaluate the stress limitations

- Bending line accessibility (i.e. stud profile before closing) – side roll passes?

- Over-bend requirements – Re: spring back effect

- Special considerations – false bend, speed differentials, tolerances

- Angle adjustment

There are two ways to create the flower:

- From fully formed shape, start to unfold the part until it is flat

- From flat to fully formed shape

4.12.2 *Flower Diagram*

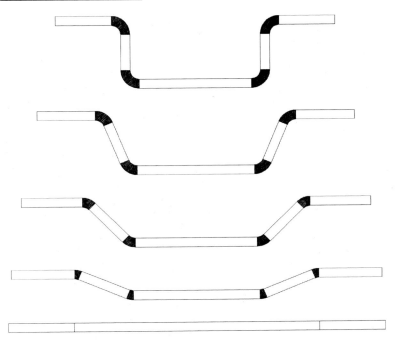

4.12.3 *Plan and side view*

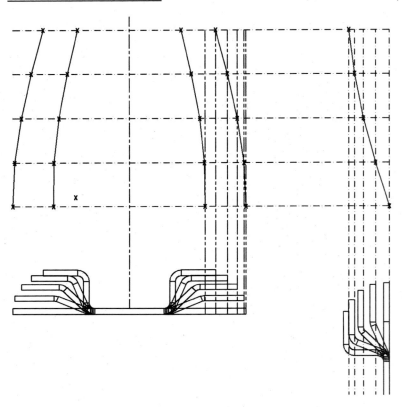

Fig. 4.12.3

The above diagram Fig. 4.12.3 shows how the edges are moving in 3D curve. If the elevation is too aggressive and the horizontal distance (forming station to forming station) is too close, the stretch will be excessive and wavy edge will occur.

4.13 Choosing Drive Line

One of the critical decisions is to choose drive diameters – see Fig. 4.13.

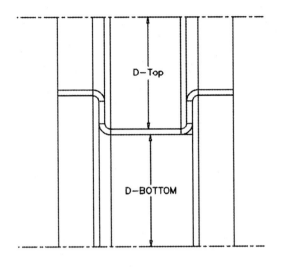

Fig. 4.13

Rule: The ratio Bottom roll diameter (Db) to Top roll diameter (Dt) must be the same as the drive Ratio – bottom shaft to top shaft.

There are 3 options:

1. To choose ratio Db:Dt as per existing roll forming machine ratio

2. Based on your Db/Dt ratio to manufacture Roll forming machine with the same ratio

3. If your chosen Db/Dt is not the same as the Roll Forming machine, what is happening is this: Let's say you choose your drive line to be with Top Diameter - 8" and your Bottom Diameter 3.75" for a Roll Forming machine that is with ratio 1:1 (i.e. Gear head type Roll forming machine). In 1 rotation Top roll will travel perimeter length of 25.13" and Bottom Roll will travel 11.78" which creates slippage on the material – for the roll with bigger diameter. As a result there will be:

- Scratches on the material especially at drive line location.

- Your cut to length tolerances will be in severe deviations

In case like that is recommended to disengage Top shaft drive gear.

Ratios for Rolls are chosen mostly for parts with deep cross section. The parts are with high legs and, if the ratio bottom to top shaft is equal (1:1), all bottom rolls have to be designed with the same pitch rolls diameters as the top shafts. This set of roll tooling will consume more rolls weight, and manufacturing is more expensive. For example "C" profiles with legs up to 4" are manufactured in roll forming machines with Ratio bottom to top shaft 1.4:1.

4.14 Features in Rolls design

4.14.1 Total Rolls length

Based on the Roll tooling space available on the shaft, the rolls should take the minimum necessary length possible. It is more economical. The rest of the available roll tooling space is fulfilled with spacers (Fig. 4.14.1).

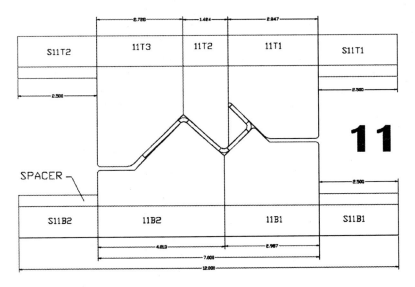

Fig. 4.14.1

4.14.2 Rolls split

Rolls are split for machining, grinding, heat treating, and loading purposes. It is recommended for 6", 8", or 10" diameter, the rolls length not to exceed 4" in length in. It is difficult for the rollforming operator to handle rolls that exceed 50lb.

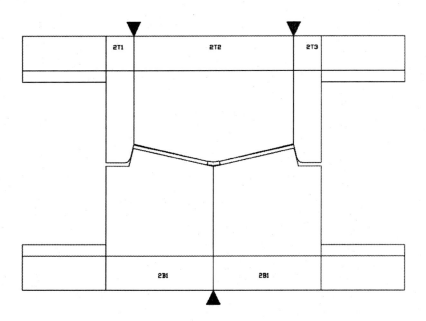

Fig. 4.14.2

4.15 Strip thickness

Maximum material strip thickness has to be considered the nominal size plus the tolerances given by the metallurgy supplier for every specific metal. If the rolls are designed and the roll tooling is set up on the nominal size, and it is supplied on the plus side, this will cause issues as a thicker material has to go through a smaller clearance. The shafts will deflect in a long run and bearings overloaded. When embossing or marking on the material is required, rolls will need groves to clear.

4.16 Strip width

Maximum material strip width has to be considered: the one calculated plus

- Strip width max tolerances – slitting standard is within +/- 0.005"

- Any grooving – add extra material or stretching

- Evaluate expected stretching at edges and leave reasonable gap space between rolls

4.17 Lead into Roll's gap from pass to pass

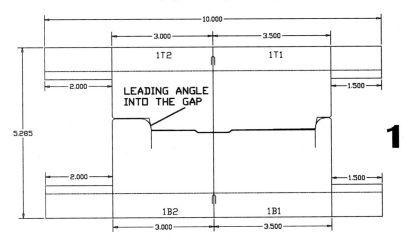

Fig. 4.17

Leading angle on the rolls helps smooth material entry between the rolls for feeding and run.

4.18 Use Gauge gap between rolls for set up purpose

The gaps are designed for gauging – one material thickness (Fig. 4.18). After set up and roll tooling tryout, the final set up gaps must be recorded on the set up sheets.

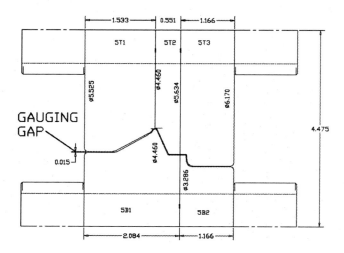

Fig. 4.18

4.19 Equal Top and Bottom Rolls total length

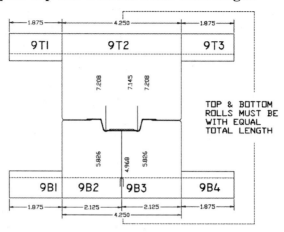

Fig. 4.19

4.20 Relieve angle

The roll that is elevating a specific segment of the part is in full contact. The opposite roll is recommended to be with relieving angle 1/2° to 2°, so that the materials can "breathe" and reduce edge stretching (Fig. 4.20).

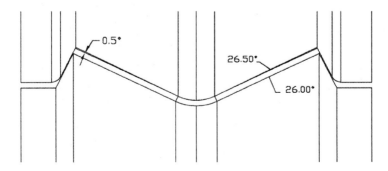

Fig. 4.20

4.21 Material trapping

Trapping is a practice of guiding and holding important dimensions within the tolerances required. Three techniques are widely used for material trapping:

a. Horizontal trap

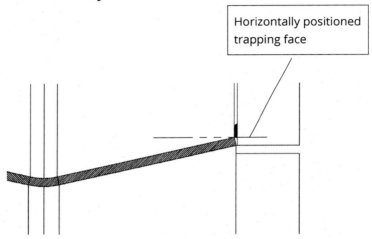

Fig. 4.21.a

b. Vertical trap

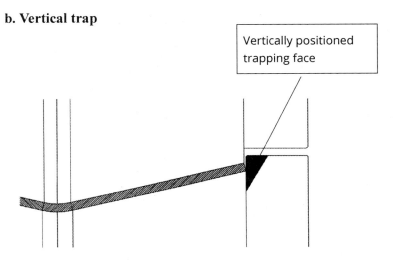

Vertically positioned trapping face

Fig. 4.21.b

c. Angular trap

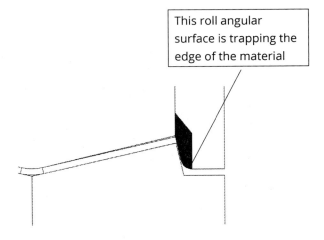

This roll angular surface is trapping the edge of the material

Fig. 4.21.c

4.22 Corners relieve

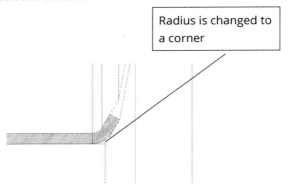

Radius is changed to a corner

Fig. 4.22

4.23 Rolls Lead

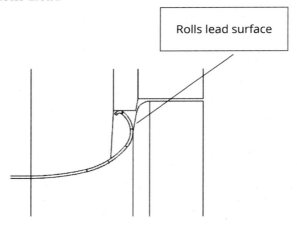

Rolls lead surface

Fig. 4.23

Dako KOLEV P.Eng.

CHAPTER 5

Trouble shooting

In the roll forming industry, it is a common knowledge that a good roll formed product (RFP) is in function of the following variables:

RFP = f (RFM, RT, RFL, RFP, D, RFPS, M, L, EC)

RFM – **Roll forming machine** – type and design features variables:

- Specific types of Rollformers are appropriate for certain types of rollformed parts. A proper match between part being rollformed and the Rollformer's design configuration and sizing is required for quality production. Often, brand new Rollformers are designed and built just to suit a specific part with all its features, especially for RFP that are with complex shape and tight tolerances.

- Horizontal distance from pass to pass – it is important for material edges stretching.

- Condition of the machine – bearing blocks condition, alignment, shafts condition (if deflected), etc.

- Quantity of Passes – there is always a dilemma if an existing Roll forming machine with certain quantity of passes will accommodate certain profile – if the machine is short on passes, the profile can still be produced, but will give you limitation in quality and speed – something will be sacrificed.

RT – **Roll tooling** with the following variables:

- Roll Tooling Design – good or bad design

- Roll tooling material – less expensive Roll tooling materials after a certain period of time start to pick up on material, wear out quickly, and decrease part quality.

- Roll tooling manufacture – questions: are all dimensions held correctly? Is the technology of manufacture completed or are certain procedures missed?

- Roll Tooling proving – in the process of roll tooling development, all coil strip width and material thicknesses must be compensated in the roll tooling.

RFL – **roll forming line configuration variables:**

- Is it a tight line?

- Is it a loose line?

- Does the line require pre-notching or punching? It is important based on the cut to length what kind of feeder is used.

- What is the cut-off system – Servo acceleration, Air cylinder, or drag with spring return?

Rollforming 101

RFP – **Roll formed product** variables:

- Profile shape – is it simple, medium complex or complex?

- RFP tolerances – loose – +/- 0.060"; medium tight – +/- 0.030"; tight – +/- 0.010"

Note:

- RFP with tight tolerances require Roll Forming Machine with more passes than usual.

- RFP with complex shape and tight tolerances will require at least 1/3 more passes and lower speed.

D – **Dies** – Pre-punching/notching and Cut-off Dies

- All pre-punching/notching operations are creating additional stress around the cut edges that later, inside of the Roll forming machine, shows tendencies of relieving that stress and works against the shape of the Roll Forming product.

RFPS – **Roll Formed product sensitivity**: The final product is always sensitive to shortage of rollforming stations and the Rollformer speed. Heat anticipation in the parts corners is causing side effects.

M – **Material** that is roll formed. It is important to understand that the final dimensions depend on:

- Material mechanical properties

- Tolerances on the material after tension leveler proceeding and slitting – all deviations in thickness work against the Roll forming machine, and all deviations in **width,**

camber and **coil set** of the strip are working against the Roll tooling (if the roll tooling can compensate those deviations)

L – **Lubrication or cooling** during rollforming:

- Based on the material being roll formed, certain lubrication is required to avoid material pick up on the rolls or for coolant to cool as much as possible the roll formed profile and avoid side effects such as bowing and twisting as a finished shape after it cools off completely.

EC - **Electrical controls:**

- Choosing proper methods of measuring controls will improve the length controls and notching/punching progressions.

- Servo Die acceleration vs. Air cylinders Die acceleration?

- Encoders vs. sensors (The encoder sooner or later will have slide slippage on the material and gives faulty readings)

- In trouble shooting, all the above variables can be narrowed down to 3 major variables in the quality formula: (1) Roll Forming machine condition, (2) Roll Tooling – inspection, set up and design review, (3) Material being Roll Formed

5.1 Rollforming machine condition

5.1.1 Rollforming Machine Inspection

Before engaging with Roll Forming Machine alignment, the following check list has to be completed:

- **Bearing blocks** – check for any axial or radial shaft's shifting. It will indicate if the bearings inside of the block are loose and need to be replaced.

- **Shafts** – Run dial indicator on the shafts – top and bottom. Bring the shafts parallel first – use grind blocks at both ends between bottom and top shaft.

- If the shaft is deflected, consider replacement.

- **Take up blocks** – clearance should be within 0.001".

- **Take up block and elevation nut** – often the elevation nut is worn out and the take up bearing block is floating up and down when the Rollformer runs. As a result, the gauging gap is not correct and the pressure applied with the top roll is lost.

5.2 Roll Forming machine Alignment

Purpose Of Alignment

To produce high quality products, the Roll Forming machines must be in excellent condition. Purpose of alignment is to establish correct DATUM line for Rolls positioning in X-Z directions.

5.2.1 DATUM LINE - Horizontal and vertical alignment

Established DATUM line runs along the face of shafts step shoulders. Both bottom and top shafts shoulders must be in one plane – Fig.5.2.1a

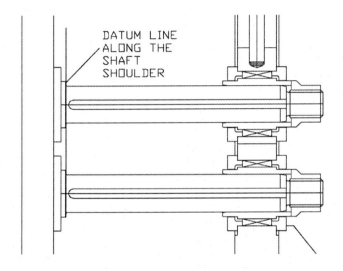

Fig. 5.2.1a

- Standard Rollformers manufacturer's tolerances are within 0.005". If one shaft is in the minus side (-) 0.005 and another is on the plus side: (+) 0.005" – the total deviation of 0.010"must be compensated. If a new set of rolls is to be loaded on an old machine, if not refurbished recently, it is almost guaranteed that due to loose bearings or worn out bearing blocks, the shaft shoulders are not inline.

- For used Roll Mills, deviation can be within 0.020"– 0.030".

- Alignment brings ALL shoulders inline horizontally and vertically.

- To bring them all inline, we measure the deviations and add shims or using alignment spacers – each one ground to the corresponding deviation.

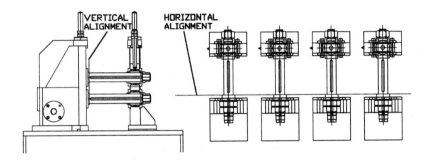

Fig. 5.2.1b

5.2.2 Methods of alignment

A. Using Musical wire and depth micrometer

B. Using straight edges and filler gauge

C. Laser alignment kit

5.2.2.A *Using Musical wire and depth micrometer*

Musical wire alignment kit Fig. 5.2.3.A is consisting of:

- Musical wire 0.024"–0.028"
- Holder with digital micrometer and light indicator
- Half spacers – Ø1 ½", Ø2", Ø3", Ø4"

Fig. 5.2.2.A

Procedure for horizontal measurement is as follows:

1. Adjust top and bottom shafts in parallel. Using caliper or grind blocks, at both ends of the shaft, and adjust inboard and outboard take up blocks.

NOTE:

- Elevate the top shafts of first and last stations 0.010" to 0.020".

Reason: the wire must not touch the shafts between; otherwise it will give wrong measuring.

2. Offset the wire from the shaft shoulder.

   - Use grind shim 1/16" to offset the wire from the shoulder.

3. Pull and lock the wire – Fig. 5.2.2.A1

   - Pull the wire between first and last pass. At both ends, use weight (Rolls from the tooling set).

   - Lock the wire by loading and packing spacers.

   - Position out board stand and tighten the locking nuts.

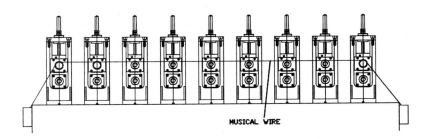

MUSICAL WIRE

Fig. 5.2.2.A1

Fig. 5.2.2.A2

Once the wire is pulled and locked in position, we start measuring the horizontal deviations with the digital micrometer holder – Fig 5.2.2.A2. The micrometer is calibrated to "Zero" by aligning it with the half shoulder edge. Slide the holder on the shaft till the half spacer touches the shaft shoulder. Turn the micrometer till the tip touches the wire and the green light goes on. Record the date on the deviation chart.

102

Measure vertical deviations

The musical wire alignment kit allows vertical measurement by using the mating half coupling and attached grind bar by positioning it on the bottom shaft and, with the micrometer, again measure the deviation from bottom to top shaft shoulder.

Another method of vertical alignment measurement is by using grind rolls with the same thickness as shown on Fig. 5.2.3.A3

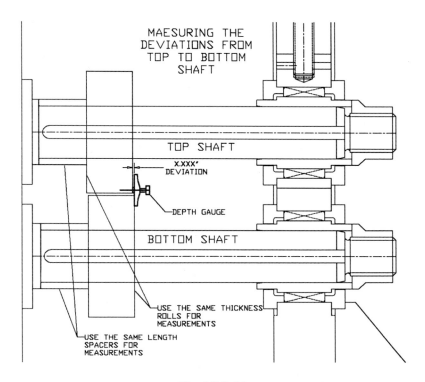

Fig. 5.2.2.A3

Deviation chart (sample)

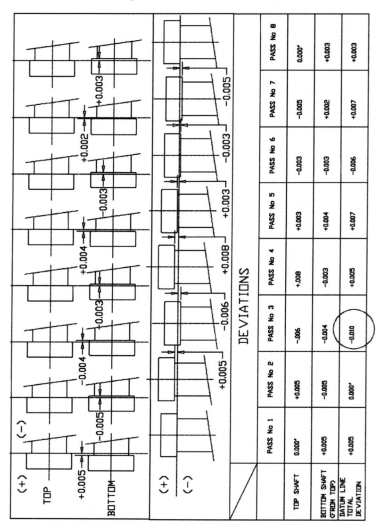

	PASS No 1	PASS No 2	PASS No 3	PASS No 4	PASS No 5	PASS No 6	PASS No 7	PASS No 8
TOP SHAFT	0.000"	+0.005	-.006	+.008	+0.003	-0.003	-0.005	0.000"
BOTTOM SHAFT (FROM TOP)	+0.005	-0.005	-0.004	-0.003	+0.004	-0.003	+0.002	+0.003
DATUM LINE TOTAL DEVIATION	+0.005	0.000"	-0.010	+0.005	+0.007	-0.006	+0.007	+0.003

Table: 5.2.2B1

All measurements taken by the micrometer are recorded in Table: 5.1.2.2B1. Simple math of adding or subtracting is showing the shaft shoulder with the most deviation.

New datum line

The shaft, which datum shoulder is the most off is:

- Bottom Pass No 3 with total deviation from the wire – 0.010"

- The new DATUM line is offset 0.010"

- Complete the correction chart – all deviations for each shaft and shims thicknesses required establishing the new DATUM line.

Calculating required shims thickness for each shaft to the new DATUM line

Correction Chart

Deviations

	PASS 1	PASS 2	PASS 3	PASS 4	PASS 5	PASS 6	PASS 7	PASS 8
T	0.000"	0.005"	-0.006"	0.008"	0.003"	-0.003"	-0.005"	0.000"
B	0.005"	-0.005"	-0.004"	-0.003"	0.004"	-0.003"	0.002"	0.003"
D=T+B	0.005"	0.000"	-0.010"	0.005"	0.007"	-0.006"	-0.003"	0.003"
Max Div			-0.010"					

Shims

	PASS 1	PASS 2	PASS 3	PASS 4	PASS 5	PASS 6	PASS 7	PASS 8
t=T-MaxDiv	0.010"	0.015"	0.004"	0.018"	0.013"	0.007"	0.005"	0.010"
b=T+B-MaxDiv	0.015"	0.010"	0.000"	0.015"	0.017"	0.004"	0.007"	0.013"

Table: 5.2.2B2

5.2.3.B Using straight edges and filler gauge

- Procedure is the same as with the musical wire; instead of wire, you can use straight edge flat bar – 10' length minimum – Fig. 5.1.2.3.B.

- Position the straight edge on the top/bottom shafts against shafts shoulder, covering minimum 5–6 stations and, using filler gauge, measure the gap between shaft shoulder and the edge.

- Use small straight edge for vertical measure as well. Preferable is a pair of flat rolls, same length.

- Fill out the chart; calculate the deviations for each shaft and shims thickness.

Fig. 5.1.2.3.B

5.2.3.C Using Laser alignment system

Laser alignment systems are available on the market for aligning Rollformers. These systems consist of a Laser, Target and Data display. Laser sends light to lines and flat planes aligned to the shaft shoulder datum and measures other points by using laser target. The accuracy is 0.00005" in 5'.

5.3 Roll Tooling – inspection, set up and design review

5.3.1 Roll tooling inspection

Roll Tooling Inspection is completed with the following steps:

- With micrometer, measure roll's diameters – flat areas, rolls length.

- Using Optical Comparator Fig. 5.3.1a – check each pair of Rolls for geometry inconsistency caused by the manufacturing process.

Fig. 5.3.1.a

- Compare with the rolls design drawings.

- Run wire gauges through the gap of each pair of Rolls (Fig 5.3.1b) to check surface consistence.

Fig. 5.3.1b

5.3.2 *Roll Tooling Set up*

When setting up a Roll Forming machine with a new set of Roll Tooling, the following steps must be performed:

- Clean up all shafts and shaft shoulder (step face).

- Clean up all rolls (Fig. 5.3.2) – any contaminations on rolls will reflect on parts appearance.

Fig. 5.3.2

- Follow up the set-up roll tooling chart – supplied by Roll Tooling manufacturer. For each pass, all rolls and spacers are numbered:

For example, as shown on Fig. 5.3.3:

Top Rolls are numbered: 3T1, 3T2, 3T3 means: Pass No 3, Top shaft, Roll #1, etc.

Bottom Rolls are numbered: 3B1, 3B2…..

Rollforming 101

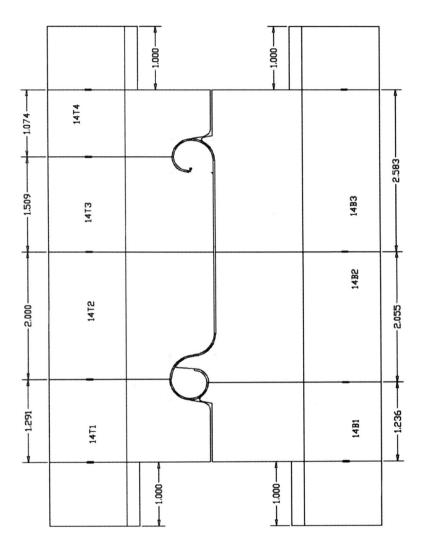

Fig. 5.3.3

- Good Roll Tooling design requires gauge shoulders of the end rolls for gap calibration.

- Using filler gauge (or a strip of actual material) – Fig. 5.3.4, adjust top to bottom rolls to correct gauge gap (material thickness). The rule is: with the material inside, the rolls should be able to slide over the material (whatever the backlash allows to check). If it is too tight, the part will come out with side effects of twist and bow.

Fig. 5.3.4

5.3.3 TRY-OUT and trouble shooting

- Run material through the Roll Forming machine and check out the finished roll formed product.

Inspect the following:

- Overall dimensions and tolerances. All general dimensions measurements must be taken 4" away from each end of the part because of flare and in the middle. Check the consistency in every 10–20 pieces.

  Unless specified, the standard tolerances in the Roll forming industry are: General dimensions: (+/-) 1/32", Bow and Camber: 1/4" per 10ft length, Length: (+/-) 1/8", Angles: +/- 2°, Twist: 5°

- Use EDM template to check if the profile fits into it.

- Check for unexpected marks and scratches on the surface or shaving edges. Rolls that are causing it, need to be modified.

5.3.4 Reverse inspection

- If your profile needs correction, you start inspection from the last forming station towards the first, and examine each cross section. Measure all dimensions till you reach the station where the problem is first generated, which roll, or pair of rolls creates the problem.

  What to look for?

- Insufficient gap adjustment between rolls. Use mirror (Fig. 5.3.4) and light reflection to check the gaps.

- Rolls are not running consistent over the material – slight deviation

Fig. 5.3.4

in Top shaft adjustment (i.e. LH is high and RH is Low). Coat the Rolls with high viscosity oil or grease and run the material through. If contact is not consistent, you will see it.

- Compressing the material too much or not enough when controlling corners

- Double tracks in the corners – happens when the rolls of the sequent forming station are not in the same track formed in the previous, but creating a new one. It happens if the Rollformer's shafts are not aligned.

5.3.5   *Found the bad one!*

- Isolating each pass, you will reach the pass where the problem is starting.

- Take out the rolls and inspect them as a pair on an Optical Comparator to see the gap consistency between the rolls.

- Run wire gauge to check for any loose spots.

- Compare with the rolls design.

5.4 Design review

If there are still any issues after completion of the above check list, further examining the cause of troubles requires new review of the rolls design. It is possible that part of the flower development is not correct, or the flower is correct, but rolls design is completed poorly – missing leading edges, guiding and material trapping, missing rolls diameters step progression etc., bending energy is not balanced or the parts have not been properly analyzed at the first stage when applying all basic principles of the rollforming technology.

Designing rolls is an art, and having a good roll tooling designer with sufficient expertise is the key for successful roll tooling completion.

After all trouble shooting analysis, you need to draw your conclusions and decide what the best course of action is to correct the rollformed profile.

- It may need specific rolls to be re-cut, certain angles relieved, or radiuses corrected. If not enough material – manufacture a new Roll.

- If it is a rolls design issue – correct the design and replace all rolls necessary.

- If the Rollformer is short of forming station, it may need additional side rolls fixtures between the rollforming stations to control the profile before entering certain pairs of rolls.

5.5 Roll Formed Material Conditions

If you did everything necessary on the rollforming machine and the roll tooling, and still have issues with the finished shape of your roll formed profile, take a look at the material supplied for rollforming.

The coil material supplied is out of:

- Tolerances in thickness and width
- Sufficient "coil set" effect

List of coil set troubles:

a. Lengthwise curvature

b. Center Buckle – humping in the center because the center is thicker and the edges are thinner

c. Crossbow – type of crosswire curvature

d. Wavy Edges – nesting in the center because the center is thinner and edges thicker.

e. Twisted strip – lengthwise bending or twisting – uneven stress distribution along the edges

When all issues are eliminated and a good part is rollformed, record on the set up sheet all final gauge gaps – inboard and outboard side for each rollforming station, keep notes on what, where, and how the trouble shooting was resolved. The relationship between a specific set of rolls and a specific rollforming machine is quite sensitive, and loading the same set of rolls on another Rollformer will have different trouble shooting challenges.

5.6 How to reduce end flare

End flare is an issue in rollforming. We cannot eliminate it, but we can reduce it. The ends deform after cutting – one end "flare-in" and the other end "flare out". End flair is caused by: residual stress in the material, stretching and compressing the edges, pre-notched edges. Flair can be reduced by increasing the quantity of rollforming stations engaged in certain rollformed product. Smooth flower and less aggressive elevation in parts segments will contribute to flair reduction. Running lubricant or coolant will reduce heat anticipation. Any notching on the cut line will contribute to the flare effect – it is recommended to avoid if possible. Additional special CAM blocks build inside of the cut-off Die can help reduce flare.

5.7 Maintenance

Proper maintenance schedules must be established and carried out by the maintenance personnel weekly, monthly, and yearly. Dailey, every rollforming operator has to inspect the rollforming line assigned to him/her for visual issues or something that disturbs the rollforming process. Good maintenance practice, at least once in every 3 months, is to:

- Inspect the rollforming machine and check all shafts, take-up blocks, bearings, and elevation nuts for any issues.

- Inspect all drive components – chain, belts, pulleys, etc.

- Grease all lubrication points.

- Change gear oil of gear reducers every 6 months (if synthetic – one year).

- Clean up the Roll Tooling. If running, for example, zinc coated material, zinc is building up on the rolls (material pick up) and must be cleaned as often as necessary. Rolls need to be taken off the machine and, in an engine lathe, be cleaned and polished with pads or fine sand paper – 160 grid.

- Align the Roll Forming machine each time when changing roll tooling.

Reference

1. Soprotivlenie materialov – P.A. Stepin

2. Machinery's Handbook

3. Roll Design – George T. Halmos – Delata Engineering Inc.

4. Engineering Mechanics – R.C. Hibbeler

Notes

Further Assistance

For assistance in finding solutions for specific rollforming application you can contact me at:

KOLEV Engineering Inc.

www.kolev.com

dako@kolev.com

Dako Kolev P.Eng.